d. Barberry.
e. Bramble.
f. Cloud Berry.

KEW POCKETBOOKS

FRUIT

Curated by Hélèna Dove and Lydia White

Kew Publishing
Royal Botanic Gardens, Kew

KEW HOLDS ONE OF THE LARGEST COLLECTIONS of botanical literature, art and archive material in the world. The library comprises 185,000 monographs and rare books, around 150,000 pamphlets, 5,000 serial titles and 25,000 maps. The Archives contain vast collections relating to Kew's long history as a global centre of plant information and a nationally important botanic garden including 7 million letters, lists, field notebooks, diaries and manuscript pages.

The Illustrations Collection comprises 200,000 watercolours, oils, prints and drawings, assembled over the last 200 years, forming an exceptional visual record of plants and fungi. Works include those of the great masters of botanical illustration such as Ehret, Redouté and the Bauer brothers, Thomas Duncanson, George Bond and Walter Hood Fitch. Our special collections include historic and contemporary originals prepared for *Curtis's Botanical Magazine*, the work of Margaret Meen, Thomas Baines, Margaret Mee, Joseph Hooker's Indian sketches, Edouard Morren's bromeliad paintings, 'Company School' works commissioned from Indian artists by Roxburgh, Wallich, Royle and others, and the Marianne North Collection, housed in the gallery named after her in Kew Gardens.

INTRODUCTION

FRUIT, WARMED BY THE SUN AND PICKED
straight from the plant, must be one of life's treasures.
Often coming in colourful packages and full of juicy
sweetness, fruits are as important to the plant as
they are to the humans and other animals which
consume them.

Botanically, the fruit of a plant is formed from the ovary
of the flower and contains the seeds, so is a crucial part
of the reproductive process. The ovary turns into a
delicious fleshy fruit after pollination has occurred, and
its main purpose is to protect the seeds within until they
are mature. Then the fruit performs its second function
of attracting animals to disperse the seeds.

Fruit offers a sweet treat to animals in the hope the
entire fruit is consumed and its seeds are carried a
distance in the animal's stomach before being deposited.
As the skin of the fruit must protect the important
package within, it can often be tough or covered in
protection such as hairs or spikes – but mammals, in
particular, have evolved to get into these skins to access
the sweet flesh inside.

Fruit is often brightly coloured, which allows it to be spotted by animals. It often starts life as green, which allows it to be well camouflaged by surrounding leaves and enables photosynthesis via chlorophyll to occur in its green skins before it ripens to a different hue. The ripe colour of fruits coincides with the maturity of its seeds, and lets animals know which fruits are ready to eat. The colour is often adapted for specific dispersers; for example, fruits attempting to attract birds are often red and black, whereas mammals tend towards brown and orange fruits. Often, the pigments which colour the skins and flesh of fruit are antioxidants, so as well as looking beautiful they do the body good.

Humans have taken full advantage of the sweet flesh of fruits, cultivating the sweetest, most colourful fruits the wild offers. These are then often bred to be even larger, sweeter and more colourful. Sometimes strategies are used to produce seedless fruits without pesky hard seeds embedded in the flesh. Such plants must be reproduced via vegetative propagation, resulting in clones of the parent plant.

Under the botanical definition, tomatoes, cucumbers, peppers, squashes and beans are all technically fruits, as they are derived from the ovary and contain seeds. But as they are predominantly consumed as savoury treats they have not made the pages of this book, which focuses on harvests used in sweet dishes.

The Royal Botanic Gardens, Kew, has a rich history concerning fruit. Even before it became a botanic garden, the area was home to several Georgian kitchen gardens containing varied fruit such as grapes, apples and plums. There were also many glasshouses dedicated to producing tropical fruits such as pineapples and peaches, which will not grow outdoors in the cold UK winters. Even today, there are areas on-site named after their previous use, which help remind staff of the importance of fruit. From the Melon Yard to the Orangery, many specific areas produced the exacting conditions needed for different crops.

During the eighteenth and nineteenth centuries, Kew received many fruits new to the country due to its role in aiding imperial expansion through plant transfers. The ackee fruit (see page 20), for example, was one of many plants introduced to Kew from the HMS *Providence* voyage (1791–1793) to transfer breadfruit (see page 40) from Tahiti to British colonies in the

Caribbean, a voyage facilitated by Joseph Banks while Kew was under his de facto directorship. Kew's large glasshouses still grow fruits from around the globe, and today Kew scientists work with the collections to help solve issues with the large-scale production of some of the world's fruits. Problems with the health of banana crops have led scientists to use the collections at Kew to help work on a solution, which could involve a pink banana-producing cousin of the banana tree, housed in the Temperate House. Accessing the collections of fruit plants at Kew helps scientists in the fight against food poverty. As well as plants, they can access fruit seeds in the Millennium Seed Bank at Kew's sister site, Wakehurst, where seeds from around the world are stored and preserved for scientific use. Current work includes banking the wild relatives of many cultivated fruits in case they hold the answer to future problems.

Growing fruit at home offers up a whole different flavour experience. Often, fruit in the supermarket is harvested before it is ripe, so that it can be transported when the skin is firmer and less likely to be damaged. Fruit cultivars in the shops are often bred for thicker skins and uniformity of size and colour, with taste coming further down the list of criteria. As fruit in the garden or allotment need not undertake long and treacherous journeys to the plate it can be harvested at the perfect time of ripeness, and cultivar choices can focus on taste. It is rare, for example, to find the heritage knobbed russet apple in shops, as it has a peculiar (some say ugly) appearance, with lumpy brown skin – but the taste is amazing.

Within these pages, fruit has been celebrated and documented by botanical artists who have captured the beauty of nature's produce. The rich colours of the skins and the often-surprising contrasting appearance of the hidden flesh has obviously inspired. These lovingly produced works are an inspiration to grow and eat many different fruits and to explore the varied taste and textures of these sweet treats.

Hélèna Dove
Head of the Kitchen Garden
Royal Botanic Gardens, Kew

Garcinia mangostana

mangosteen
from Berthe Hoola Van Nooten
*Fleurs, Fruits et Feuillages Choisis de la Flore
et de la Pomone de L'Ile de Java*, 1863

Native to areas around the Indian Ocean, the
dark red skin of the mangosteen reveals starkly
contrasting white flesh when opened. It is an
incredibly juicy fruit which has lots of juice
filled pockets and has a mild, fragrant flavour.

Cydonia oblonga

quince

from Alois Sterler and Johann Nepomuck
Mayerhoffer *Europa's Medicinische Flora*, 1820

Once an extremely popular fruit, quince has
golden skin, creamy flesh and a strong aromatic
flavour. The flesh is extremely hard when picked
in autumn, so it is mainly stewed when eaten,
often alongside apple to temper the strong
quince flavour. It is also turned into a
cheese called membrillo.

Fragaria species

strawberry

from Antoine Poiteau *Pomologie Française*, 1846

This tasty summer delight is in fact a false fruit.
Botanically, the flesh of a fruit is predominately
made up of the swollen ovary. But in the case
of the strawberry, most of the sweet red
harvest is formed of the receptacle, the base
of the flower that holds the ovary and petals,
which is embedded with hundreds
of tiny, hard true fruits.

Passiflora edulis

passion fruit, purple granadilla
from *Curtis's Botanical Magazine*, 1818

Passion fruit delivers a sharp, sweet tang to
fruit salads but, sadly, the vine it grows on
is not hardy and does not thrive outdoors in
colder climates. Thankfully, the fruits of most
Passiflora, including the commonly grown
decorative type *P. caerulea*, are also edible
and the plant is much hardier.

Blighia sapida

ackee, akee
from Michel Étienne Descourtilz
Flore Médicale des Antilles, 1821–9

Ackee is a bright red fruit which turns orange and
splits open when ripe. Inside are 2–4 chambers,
each containing a large black seed surrounded
by creamy flesh, which has a nutty taste and
is harvested to cook into savoury dishes.

Vaccinium macrocarpon

cranberry, American cranberry
from *Gartenflora*, 1871

The small, red, tart fruit of the cranberry grows on an evergreen shrub which crawls along the ground. Commercially, they are grown in bogs as they are ericaceous and thrive in the acidic pH. Some cranberries are even harvested with water: first the fruits are cut from the plant, and then the area is flooded. As cranberries contain air pockets, they float along making them easy to sort.

Averrhoa carambola

carambola, star fruit
by Marianne North from the
Marianne North Collection, Kew, 1876

The carambola fruit has five ridged edges,
meaning its cross section is the shape of a star,
leading to its other common name the star fruit.
The whole fruit is edible, with a slightly waxy
skin and crunchy flesh which has a sweet taste,
akin to a grape, although there are also sour
flavoured cultivars.

Prunus domestica

plum, European plum
from Alois Sterler and Johann Nepomuck
Mayerhoffer *Europa's Medicinische Flora*, 1820

———————

Plums grow in a vibrant array of colours, from
deep purple to golden yellow and lime green.
They also vary in flavour from tart to sweet.
All plums are covered in a waxy layer called
the wax bloom, which reduces water loss.

Diospyros kaki

persimmon, Sharon fruit
from *Revue Horticole*, 1878

The orange fruit of the persimmon tree is often called a Sharon fruit, or simply a persimmon. It is entirely edible including the skin, and delicious once ripe, although tough and astringent before this. The tree is hardy, meaning the persimmon can be grown in colder climates, although fruit may need to be ripened indoors.

Ribes uva-crispa

gooseberry

from George Brookshaw *Pomona Britannica*, 1817

Growing on a small shrub, gooseberries hang
like little jewels, which promise a burst of
flavour either sweet or sharp depending on
the cultivar. The shrub is covered in sharp
spines, making harvesting the berries slightly
treacherous, but more rewarding.

Crataegus germanica

medlar

from *La Belgique Horticole*, 1856

The medlar is a small brown fruit with
triangular calyx lobes which hang down,
giving an unusual appearance and leading to
nicknames such as the donkey's bottom. The
fruit is only soft enough to eat after it has
been bletted, which means leaving it on
the tree long enough for a frost to
break down the tough flesh.

Citrus x *limon*

lemon

from Antoine Poiteau *Pomologie Française*, 1846

Lemons grow on evergreen shrubs that do not tolerate cold winters, so in the UK they need to be moved indoors when a chill is in the air. They tend to flower in later winter, and as the fruits take around 12 months to ripen they will have beautifully scented flowers and stunning yellow lemons at the same time.

Mangifera indica

mango
from Berthe Hoola Van Nooten
*Fleurs, Fruits et Feuillages Choisis de la Flore
et de la Pomone de L'Ile de Java*, 1863

Mangos grow on large trees in warm, tropical
climates. The fruit's sweet juicy flesh surrounds
a large stone which protects the seeds. In the
wild, mangos are predominantly pollinated by
fruit bats, who also consume the ripe fruits.

Momordica charantia

bitter melon, bitter gourd, balsam pear
from *Flore des serres et des jardins de l'Europe*, 1854

———————

Like cucumbers and squashes, bitter melons
belong to the Cucurbitaceae family and grow
on a trailing vine. Fruits are usually harvested
when immature and still green. The skin of
bitter melons has a warty texture, and the fruit
is hollow in cross-section. They are sliced and
cooked as a vegetable, which retains a
distinctive bitter flavour.

Artocarpus altilis

breadfruit

from Joseph Jacob Ritter von Plenck
Icones Plantarum Medicinalium, 1788–1812

Breadfruit is composed of hundreds of smaller
fruits all fused together, which can easily be
seen from the hexagonal markings on the skin.
It can be eaten raw, but is more commonly
cooked, where its predominantly starchy flesh
takes on the consistency and taste of bread,
hence the common name.

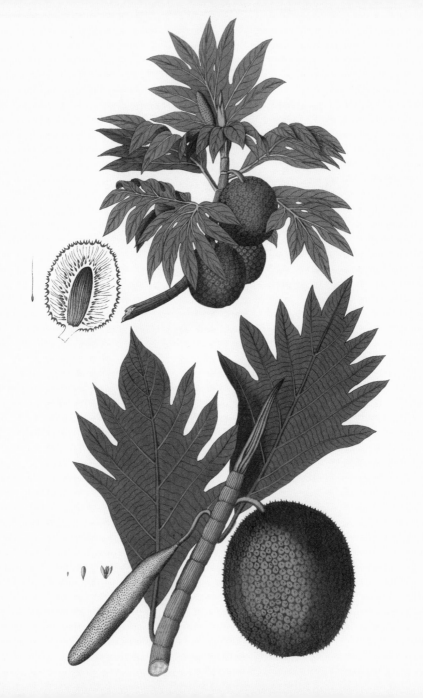

Manilkara zapota

sapodilla, beef apple, chikoo
from Jean Claude Mien Mordant de Launay
Herbier Général de l'Amateur, 1816–27

The sapodilla fruit grows wild in warm climates
such as Mexico, and can get to heights of 30 m
(98 ft), although in cultivation they are kept at
manageable heights of around 15 m (49 ft).
The flesh of the fruit varies from creamy to
chocolate brown, and once ripe is soft and
sweet with a rich malt quality.

Musa x *paradisiaca*

banana, French plantain
from Berthe Hoola Van Nooten
*Fleurs, Fruits et Feuillages Choisis de la Flore
et de la Pomone de L'Ile de Java*, 1863

Bananas are thought to have been the first
fruit on earth. Commercial bananas are
seedless, which makes them pleasant to eat
but means that the plants must be propagated
vegetatively, so all bananas are identical clones.
Banana plants grow from a rhizome and their
appearance is tree like, but actually they are
herbaceous with a very tall central stem.

Nephelium lappaceum

rambutan

from Berthe Hoola Van Nooten

*Fleurs, Fruits et Feuillages Choisis de la Flore
et de la Pomone de L'Ile de Java,* 1863

The fruits of the rambutan tree grow in clusters
and have a skin covered in fleshy, red spines
which are also known as spinterns, giving fruit
a hairy appearance. The tree itself is sensitive
to the cold and struggles to grow in
temperatures below 10°C (50°F), limiting
the areas in which this fruit will thrive.

Pyrus species

pear

from Antoine Poiteau *Pomologie Française*, 1846

Another member of the quintessential orchard, the pear is a sweet fruit which is best eaten straight off the tree as its thin skin does not travel well. One notable variety is Williams Bon Chrétien, a cooking and eating type, which, the story goes, was named by W. T. Aiton who managed the Gardens at Kew, including the Kitchen Garden.

Alkekengi officinarum

Chinese lantern, physalis, bladder cherry
from *Flore des serres et des jardins de l'Europe*, 1854

Mainly grown for the stunning papery husks
that protect the fruit whilst it develops, and
resemble lanterns hanging from the stems. The
fruits of this species are fairly tasteless and
sometimes used in medicines. Its close relative,
the Cape gooseberry, *Physalis peruviana*, has
much more succulent, sweet golden fruits, but
sadly the husks are less colourful.

Nº 21

Annona mucosa

lemon meringue pie fruit
from François Richard de Tussac
Flore des Antilles, 1808–27

This large tropical tree bears giant fruits,
covered in a layer of soft spines. Once opened,
the flesh tastes like lemon meringue pie, sweet
and fragrant. Unfortunately, the fruit bruises
easily and does not store well once harvested,
so is a treat best freshly picked.

Litchi chinensis

lychee, litchee, alligator strawberry
from Jean Claude Mien Mordant de Launay
Herbier Général de l'Amateur, 1816–27

––––––––––––––

The lychee fruit originates in China and grows
on an evergreen tree of around 15 m (49 ft).
Harvested for their sweet flesh, which is eaten
raw, the fruit is surrounded by a distinctive,
textured, red skin, which gives them their other
common name of alligator strawberry.

Citrus x *aurantium*

**Seville orange, bigarade, bitter orange,
marmalade orange, sour orange**
from Antoine Risso, Antoine Poiteau and
Alphonse Du Breuil *Histoire et Culture
des Orangers,* 1782

———————

The fruit of the orange comes wrapped in its
own protective casing, which is commonly
called peel and makes a delicious ingredient in
its own right. The white side of the peel is
called the albedo; the orange side, the flavedo,
contains most of the volatile oils which gives
the peel its flavour.

Syzygium cumini

jambolan, **Java plum**, **black plum**
from Berthe Hoola Van Nooten
*Fleurs, Fruits et Feuillages Choisis de la Flore
et de la Pomone de L'Ile de Java*, 1863

———

The fruits of the jambolan tree ripen in colour
from immature green through to coral pink
and then to a deep black once ready to harvest.
The taste can be sweet but is often quite acidic
with astringent properties – but as they are
incredibly juicy, they are useful in
jams and preserves.

Eriobotrya japonica

loquat

by Jean Gabriel Prêtre from the Kew Collection, 1825

Loquats are small, orange fruits with white, tart
flesh, which flower in autumn and winter and
are ready to harvest in the spring. The seeds take
up a lot of room within the fruit, so it is common
to cook the fruit and strain the seeds. The flesh
contains high amounts of pectin, so it is great
for setting preserves.

Rheum officinale

rhubarb, Chinese rhubarb, medicinal rhubarb
from Charlotte Mary Yonge *The Instructive Picture Book or Lessons from the Vegetable World*, 1863

Although technically not a fruit but a vegetable, as its edible leaf stalks are part of the vegetative area of the plant, rhubarb's vivid pink stems are mostly eaten as a fruit in sweet dishes. It is a rarity in the kitchen garden as it is a perennial crop alongside others such as asparagus and globe artichokes. Rhubarb can be forced in early spring, by excluding light, which gives incredibly sweet and delicate stems.

Prunus avium

cherry, sweet cherry
from *La Belgique Horticole*, 1853

There are two main categories of cherry
cultivated for harvesting: the sweet cherry,
Prunus avium, which includes favourites such
as 'Stella', and the sour cherry, *Prunus cerasus*,
whose tart fruits are perfect for cooking. As well
as these beautiful red, yellow or white fruits,
cherry trees are celebrated for their beautiful
blossom in spring.

Carica papaya

papaya, pawpaw
by Georg Dionysius Ehret from
Christoph Jacob Trew *Plantae Selectae*, 1750–73

Papaya fruits grow up the massive central stem
of this herbaceous plant, which can get to
5 m (16 ft) and above. The plant often sheds
the lower leaves, and ends up looking like a
large, tropical Brussels sprout plant. The
fruits are generally harvested when the flesh
is sweet and deep orange, but they can be
picked when immature and green to be
cooked in savoury dishes.

Vitis vinifera

grape vine, cultivated grape
from Antoine Poiteau *Pomologie Française*, 1846

Grapes grow in a plethora of colours, forming
bunches on a woody vine. They have been bred
in cultivation for two purposes: table grapes for
eating raw, which are large, seedless and have
thin skins, and wine grapes, which are smaller
and often contain seeds. Wine grapes have thick
skins, which contain the heady wine aromas and
their sweetness aids the fermentation process.

Feijoa sellowiana

pineapple guava, feijoa
from *Curtis's Botanical Magazine*, 1898

The pineapple guava is not a true guava, but has
an equally fragrant and delicious flesh. It grows
on an evergreen shrub and produces beautiful
pink flowers, which also taste stunning and are
regularly used in the restaurants at Kew.

Rubus idaeus

raspberry
from George Brookshaw *Pomona Britannica*, 1817

There are two types of raspberries in common
cultivation, summer or autumn fruiting.
Growing both extends the season of production,
but they are pruned differently. Summer
fruiting raspberries produce fruit on previous
season's growth, so, at the end of summer, only
fruited canes should be cut to the ground,
leaving this season's canes. Autumn fruiting
raspberries fruit on current season's canes,
so in late winter all are cut down.

Persea americana

avocado, alligator pear
by Marianne North from the
Marianne North Collection, Kew, 1880

———————

An ever popular fruit, avocado's green flesh
is commonly consumed in savoury dishes.
Avocado trees take six years to complete the
juvenile stage. Once mature they produce pear
shaped fruits which require warm conditions
to ripen, although the Hass avocado is showing
some tolerance to cooler climates.

Punica granatum

pomegranate, granada
from Joseph Jacob Ritter von Plenck
Icones Plantarum Medicinalium, 1788–1812

Pomegranates are relatively large, ruby
red fruits full of seeds. Each of the seeds is
surrounded by a seed coat which is full of
delicious juice. The juice itself is fairly sour
with a pH of 4.4, which adds a zesty zing to
a fruit salad.

Ficus carica

fig

from Otto Wilhelm Thomé *Flora von Deutschland Österreich und der Schweiz*, 1886–9

A fresh fig is an absolute luxury, with sweet flesh and edible dark purple skin. The fig itself is developed from a series of flowers which are never visible to the human eye, as they grow and are pollinated on the inside of the fig pod or syconium.

Cucumis melo

melon
by Elizabeth Blackwell from Elizabeth Blackwell
A Curious Herbal, 1737

―――――――

Melons come in all shapes and sizes, from green
to white to orange fleshed, and the taste can be
refreshing or incredibly sweet and aromatic.
They vary in size and shape and the skin goes
from completely smooth to bearing raised
netting. The plant itself is a climber, producing
tendrils that grab on to objects to help haul
the melons up to the sunlight.

Malus domestica

apple

by E. D. Smith from Charles McIntosh
Flora and Pomona, 1829

Apples are synonymous with British orchards,
although they originate in Kazakhstan and
reached the UK via the Silk Roads around
10,000 years ago. Apples grow in an array of
colours from acid green to deep red and russet
brown. There are thousands of cultivars,
broadly classified by taste as sweet eaters, tart
cookers or bitter cider apples.

Cullenia species

durian
from Berthe Hoola Van Nooten
*Fleurs, Fruits et Feuillages Choisis de la Flore
et de la Pomone de L'Ile de Java*, 1863

The durian is proof that not all fruit is sweet. Its
thorn-covered skin should be a telltale sign to
what lies within, as the custardy flesh has such a
strong unpleasant odour that is banned in some
public places. But still, some say the smell is
sweet, and the flesh is a delicacy to some.

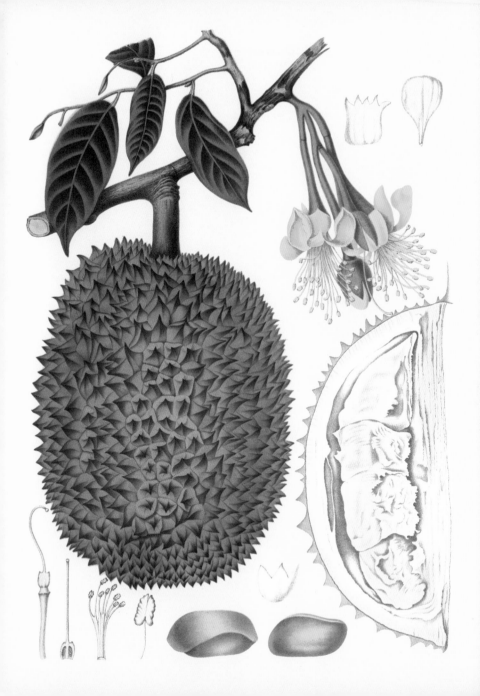

Prunus armeniaca

apricot

by J. Guillot from *Revue Horticole*, 1910

Apricot trees produce beautiful pink blossoms extremely early in the spring, before many pollinating insects are active. To ensure successful pollination, growers often hand pollinate by moving pollen from one flower to the next. Traditionally, this was done with a rabbit's tail on a long pole; a paintbrush also completes the task successfully.

Ribes rubrum, Ribes nigrum

redcurrant, blackcurrant
from *Gartenflora*, 1867

The deep purple fruits of the blackcurrant
shrub hang down in little trusses, known as
strigs, which are also characteristic of its close
relatives the redcurrant and white currant.
Due to their high vitamin C content,
blackcurrants were a popular crop during
the Second World War.

Ananas comosus

pineapple
by P. de Pannemaeker after a drawing by
C. T. Rosenberg, Kew Collection c. 1850

This tropical fruit sits atop a base of extremely
spiky leaves, which keep it well defended from
predators. The fruit itself is comprised of
hundreds of smaller berry type fruits, which
fuse together to form one large, delicious and
juicy pineapple.

ILLUSTRATION SOURCES

Books and Journals

Andre, O. (1900). *Persea americana. Revue Horticole.* Volume 72.

André, O. and Bois, D. (1910). *Prunus armeniaca. Revue Horticole.* Volume 82.

Blackwell, E. (1737). *A Curious Herbal.* Volume II. J. Nourse, London.

Brookshaw, G. (1817). *Pomona Britannica.* Bensley and Son, London.

Carrière, E. A. (1878). *Diospyros lycopersicon. Revue Horticole.* Volume 50.

Descourtilz, M. É. (1821–9). *Flore Pittoresque et Médicale des Antilles.* Volume VIII. Chez Corsnier, Paris.

Hooker, J. D. (1898). *Feijoa sellowiana. Curtis's Botanical Magazine.* Volume 124, t. 7620.

McIntosh, C. (1829). *Flora and Pomona.* Thomas Kelly, London.

Mordant de Launay, J. C. M. (1816–27). *Herbier Général de l'Amateur.* Volume VI and VII. Libraire Audot, Paris.

Morren, C. (1853). *Prunus cerasus. La Belgique Horticole.* Volume 3, p. 65.

Morren, C. (1856). *Mespilus germanica. La Belgique Horticole.* Volume 6, t. 63.

Plenck, J. J. R. von (1788–1812). *Icones Plantarum Medicinalium.* Vienna.

Poiteau, A. (1846). *Pomologie Française.* Volume II and IV. Langlois and Leclercq, Paris.

Regel, E.A. von (1867). *Ribes nigrum, Ribes nigrum. Gartenflora.* Volume 16, t. 16.

Regel, E.A. von (1871). *Oxycoccus macrocarpus. Gartenflora.* Volume 20, t. 673.

Risso, A., Poiteau. A. and Du Breuil, A. (1782). *Histoire et Culture des Orangers.* Henri Plon and Co., Paris.

Sims, J. (1818). *Passiflora edulis. Curtis's Botanical Magazine.* Volume 45, t. 1989.

Sterler, A. and Mayerhoffer, J. N. (1820). *Europa's Medicinische Flora*. J. H. Mayerhoffer, Munich.

Thomé, O. W. (1886–9). *Flora von Deutschland Österreich und der Schweiz*. Volume II. F. E. Köhler, Gera-Untermhaus.

Trew, C. J. (1750–73). *Plantae Selectae*. Volume VII. Nuremberg.

Tussac, F. R. de (1808–27). *Flore des Antilles*. Apud auctorem et F. Schoell, Paris.

Van Houtte, L. (1854). *Physalis alkekengi. Flore des serres et des jardin de l'Europe*. Volume 10, t. 1010.

Van Houtte, L. (1854). *Momordica charantia. Flore des serres et des jardin de l'Europe*. Volume 10, t. 1047.

Van Nooten, B. H. (1863). *Fleurs, Fruits et Feuillages Choisis de la Flore et de la Pomone de L'Ile de Java*. Emile Tarlier, Brussels.

Yonge, C. M. (1863). *The Instructive Picture Book or Lessons from the Vegetable World*. Edmonston and Douglas, Edinburgh.

Art Collections

Marianne North (1830–90). Comprising over 800 oils on paper, showing plants in their natural settings, painted by North, who recorded the world's flora during travels from 1871 to 1885, with visits to 16 countries in 5 continents. The main collection is on display in the Marianne North Gallery at Kew Gardens, bequeathed by North and built according to her instructions, first opened in 1882.

ACKNOWLEDGEMENTS

Kew Publishing would like to thank the following for their help with this publication: in Kew's Library and Archives Fiona Ainsworth, Craig Brough, Rosie Eddisford and Anne Marshall; for digitisation work, Paul Little.

FURTHER READING

Baker, Harry. (1998). *The Fruit Garden Displayed* 8th edition, revised. The Royal Horticultural Society, London.

Fry, Carolyn. (2013). *Kew's Global Kitchen Cookbook*. Royal Botanic Gardens, Kew.

Linford, Jenny. (2022). *The Kew Gardens Cookbook*. Royal Botanic Gardens, Kew.

Maguire, Kay. (2019). *The Kew Gardener's Guide to Growing Fruit*. White Lion Publishing, London in association with the Royal Botanic Gardens, Kew.

Mills, Christopher. (2016). *The Botanical Treasury*. Welbeck Publishing, London in association with the Royal Botanic Gardens, Kew.

North, Marianne and Mills, Christopher. (2018). *Marianne North: The Kew Collection*. Royal Botanic Gardens, Kew.

Payne, Michelle. (2016). *Marianne North: A Very Intrepid Painter*, revised edition. Royal Botanic Gardens, Kew.

Pike, Ben. (2011). *The Fruit Tree Handbook*. Green Books, Totnes.

Vaughan, J. G. and Geissler, C. A. (2009). *The New Oxford Book of Food Plants*. Oxford University Press, Oxford.

van Wyk, Ben-Erik. (2019). *Food Plants of the World*. CABI, Wallingford.

INDEX

First published in 2022
Royal Botanic Gardens, Kew,
Richmond, Surrey, TW9 3AB, UK
www.kew.org

ISBN 978 1 84246 752 7

Distributed on behalf of the Royal Botanic Gardens, Kew in North America by the University of Chicago Press, 1427 East 60th St, Chicago, IL 60637, USA.

British Library Cataloguing in Publication Data
A catalogue record for this book is available from the British Library

Design: Ocky Murray
Page layout: Christine Beard
Image work: Nicola Erdpresser
Production Manager: Jo Pillai
Copy-editing: Michelle Payne
Proofreading: Ruth Linklater

Printed and bound in Italy by Printer Trento srl.

Front cover image: *Musa* x *paradisiaca* (see page 44)

Endpapers: Fruits from Charlotte Mary Yonge *The Instructive Picture Book or Lessons from the Vegetable World*, 1863

p2: *Fragaria* species, strawberry, from *Revue Horticole*, 1900

p4: *Persea americana*, avocado, from *Revue Horticole*, 1900

p10–11: *Averrhoa* species, star fruit, by unknown artist from the Kerr Collection, Kew, c. 1803

For information or to purchase all Kew titles please visit shop.kew.org/kewbooksonline or email publishing@kew.org

Kew's mission is to understand and protect plants and fungi, for the wellbeing of people and the future of all life on Earth.

Kew receives approximately one third of its funding from Government through the Department for Environment, Food and Rural Affairs (Defra). All other funding needed to support Kew's vital work comes from members, foundations, donors and commercial activities, including book sales.

Publisher's note about names
The scientific names of the plants featured in this book are current, Kew accepted names at the time of going to press. They may differ from those used in original-source publications. The common names given are those most often used in the English language, or sometimes vernacular names used for the plants in their native countries.